HOW TO GROW TOP QUALITY CORN

Dr. Harold Willis

A Biological Farmer's Guide

HOW TO GROW TOP QUALITY CORN

Dr. Harold Willis

A Biological Farmer's Guide

Acres U.S.A.
Austin, TX

How to Grow Top Quality Corn

2nd edition, revised
Copyright © 1984, 2009 by Harold L. Willis

Drawings and photographs are by the author.

The information in this book is true and complete to the best of
our knowledge. All recommendations are made without guarantee
on the part of the author and Acres U.S.A. The author and publisher
disclaim any liability in connection with the use or misuse of this
information.

Acres U.S.A.
P.O. Box 91299
Austin, Texas 78709 U.S.A.
(512) 892-4400 • fax (512) 892-4448
info@acresusa.com • www.acresusa.com

Printed in the United States of America

Publisher's Cataloging-in-Publication

Willis, Harold L., 1940-
How to grow top quality corn / Harold L. Willis. Austin, TX,
ACRES U.S.A., 2009
    viii, 72 pp., 23 cm.
    Includes Index
    Includes Bibliography
    Includes Illustrations
    ISBN 978-1-60173-014-5 (trade)

SB205.C4 W55 2009   633.15

CONTENTS

FOREWORD

Modern agriculture is sort of like one of those two-faced masks that symbolize the theater; one side happy, the other side sad. We get the happy side from the glossy farm magazines and from the media propaganda designed for the consumption of politicians and city folk. We hear what a great job the American farmer is doing—keeping the world from starvation, with yields high. Thanks to modern science and technology, the future is rosy.

But as almost any farmer knows, there is the other side of the mask. While yields climb, quality goes down. How many people grow "standard" 56 lb. test weight corn these days? Isn't 52, 50, 48, or 46 lb. corn all too common? Then there are increasing health problems among the animals that eat our crops, some so difficult that veterinarians throw up their hands in despair. Farm income is unpredictable, rarely adequate and usually too low, while costs, inflation, and everyone else's income are in high gear. Farm bankruptcies and foreclosures soar. And on and on.

Modern agriculture is in trouble! Most people farming today won't be in 5 or 10 years, if present trends continue. Farming has to be a paying proposition—that is, the farmer has to be paid a fair profit as are other segments of the economy. Until such changes come about, one way to "beat the system" is to grow higher quality crops with less dollar input. Crops that command premium prices on the market, or when fed to your animals, produce healthy, high-producing animals.

Believe it or not, many of our current methods of growing crops will nearly always produce poor quality "foodless food." We use fertilizers and other agricultural chemicals that kill the life in the soil, which if allowed to live would help us grow good food. Soil becomes hard and tight—sterile. Weed and pest problems grow worse.

What can you do? You can begin to put your soil back into good condition by stopping harmful practices and starting right ones. You can grow top quality corn and other crops.

This book will get you started.

CHAPTER 1

About Corn—
Facts and Controversies

Corn, commonly called maize in much of the world, is without doubt America's most valuable agricultural crop. The U.S.A. produces nearly half of the world's corn. Among field crops, world corn acreage is second only to wheat. Corn can be grown in every state except Alaska, but about 80% of U.S. corn is grown in the "corn belt," which includes all or part of Iowa, Illinois, Indiana, Ohio, Minnesota, Wisconsin, Missouri, Nebraska, and South Dakota. Iowa and Illinois produce roughly 40% of U.S. corn.

The great majority of U.S. corn is fed to livestock, both as grain and silage or fodder. Beef and dairy cattle, hogs, sheep, and poultry are the major consumers. Less than 10% is used for human consumption, but in recent years the production of corn-based ethanol has used up a sizeable proportion of corn production. Besides such familiar products as corn flakes, popcorn, corn meal, hominy, grits, corn starch, corn oil, corn syrup, and corncob pipes, other industrial uses include the production of chemicals (alcohols and liquors, acetone, furfural, antibiotics, enzymes, organic acids), plastics, paper, cardboard, insulation, grit for polishing, carrier for pesticides and fertilizers, animal litter, hand soaps, and cosmetics.

History. The history of domesticated corn has been traced back several thousand years. It is believed to have originated in Mexico from a wild ancestor or from crossing with a grass-like plant called teosinte, but the details of its origin are uncertain and still debated by botanists. Corn was the basis of the ancient Aztec, Maya, and Inca civilizations. Corn was introduced to the first European settlers of America by the native Indians. The story of how the Indians taught the Jamestown settlers to place a fish in each hill of corn for fertilizer is known by every schoolchild.

Pioneering farmers and later generations took native varieties of corn and selected desirable kernels to propagate. Soon, hundreds of local varieties of open pollinated corn were produced, each well suited to its local climate and soil. In 1905, two researchers, G. H. Shull and E. M. East, independently began studying corn inheritance and produced the first modern hybrids. (However, it is believed that hybridization was carried on by Indians, and some early farmers in the 1700s and 1800s produced new varieties.) The first commercial hybrids were introduced in 1933, and since the late 1930s, hybrids have taken over almost all U.S. cornland, and are rapidly displacing open pollinated varieties in most foreign countries. And of course the development of genetically engineered or GM corn has dramatically altered the corn growing scene since the mid-1990s.

Along with the introduction of hybrid varieties, the increased use of commercial fertilizers, herbicides, and insecticides in the last several decades has caused corn yields to skyrocket:

AVERAGE YIELDS FROM USDA STATISTICS, BU/ACRE

1931-35	22.8
1941-45	32.8
1951-55	39.1
1957	47.1
1960	54.7
1966	73.1
1969	85.9
1973	91.3
1979	109.7
1985	118.0
1990	118.5
2000	136.9

What is corn? Technically corn is called *Zea mays,* and is a member of the large plant family, the grasses. If corn seems too large to be a grass, remember that sugar cane and bamboo are even larger grasses. The grass family is characterized by a slender, jointed stalk and narrow leaves. Grass flowers are small and without petals, and are generally pollinated by the wind. In corn, the flowers are divided, with the male flowers (tassel) at the top and female (cob and silks) in the center of the plant. The grain, or "seed" (technically it is a fruit) is then produced. Grasses are the source of most of the food eaten by man and livestock, since they include corn and the cereal grains (wheat, rice, oats, etc.), as well as forage grasses.

The life of a corn plant. A corn plant is a marvel of energy production and storage. Basically, the plant is an "energy factory," capturing the sun's energy and by the process of photosynthesis, converting it into energy stored in food molecules. In only 3 or 4 months, a single kernel of corn grows to a plant 7 to 12 feet high and produces 600 to 1000 kernels similar to the one that was planted.

Let's follow a corn plant through its busy life and see what goes on. There are six stages in the life of a corn plant, some of which are very critical in determining survival and yield.

1. *Germination and seedling establishment.* Before it germinates, the kernel consists of (a) the outer, hard, protective *seed coat* (pericarp), (b) the large *endosperm,* a source of stored food (about 90% starch, 7% protein, and small amounts of oil, minerals, etc.), and (c) the *embryo,* the undeveloped new plant. Upon planting, the kernel absorbs water and swells, activating the metabolic "machinery" of the living embryo and endosperm cells.

If the temperature is warm, the embryonic root (radicle) emerges from the seed coat after two or three days and rapidly grows out into the soil. Next the embryonic leaf (plumule) emerges and grows upward. Other roots branch out and the first leaf comes out of the ground in about six to ten days. The growing plant has been living off of food stored in the endosperm. After the leaves are in sunlight, the plant switches over to making its own food by photosynthesis. Rapid growth occurs for one to one and one-half weeks, with a new leaf produced about every three days.

2. *Vegetative growth.* For another two weeks, additional roots and leaves are formed so that large amounts of food can be produced by photosynthesis. The new roots grow out of the bottom of the stalk (the "crown"), which is below the surface. Later, additional prop or brace roots will grow out from above-ground parts of the stalk. The total number of leaves, 20 or more, is already formed in the growing tip in the whorl before the plant is knee high.

3. *Flower (tassel and ear) formation.* After about 30 days from germination, when the plant is about knee high, tiny tassel and ear buds are formed, when about eight to ten leaves have unfolded. Just after this, the plant grows rapidly upward and unfolds all its leaves. Tassel and ear buds enlarge and develop pollen grains and silks. The tassel emerges at the top of the plant.

This stage is the most critical in the life of the corn plant with regard to reproduction (grain yield). Large amounts of water and nutrients, and good weather are necessary to ensure large numbers of healthy kernels and pollen.

4. *Flowering (pollen shedding and silking).* One or two days after the tassel emerges, its hundreds of "pollen sacs" (anthers) burst open and shed millions of pollen grains. The ear contains about 750–1000 female flowers, each containing a potential kernel (ovule) and a long silk. The silks emerge two to three days after the pollen is shed and can receive pollen for three to five days.

After the silks are pollinated the fertilized ovules begin to develop into new kernels. This stage is also very critical; delayed silking or poor weather can greatly decrease yield.

5. *Kernel development and maturation.* After fertilization, the silks wilt and turn brown. In about two weeks the tiny kernels swell to full size as they are filled with a milky fluid containing much sugar. Then the sugars are gradually converted into starches and the kernels become harder. By the end of the 7th week after pollination, food storage slows and the kernel and embryo are nearly mature.

6. *Maturity and drying.* By the end of the 8th week after pollination, the kernels have reached their maximum dry weight and are considered physiologically mature. But before they can be harvested and stored, they must lose considerable moisture, drying from about 30% moisture to below 15% in average corn (high quality corn will store at higher moisture). Some moisture evaporates directly through the seed coat and husk, but much also is carried out through the stalk. Eventually, the whole plant dies and dries out.

Controversial choices. Before getting into the nuts and bolts of growing top quality corn, let's discuss a few subjects that relate to that goal and which may be controversial in some circles.

Rotation or continuous corn? Most extension agents and agronomists tell us that it is best to rotate a corn crop with small grains and hay crops because continuous corn tends to deplete soil organic matter and degrade soil structure. Yet some experts say continuous corn can be grown, and many farmers are doing it. Which is best?

Many studies have shown that yields in continuous corn often steadily decrease and that they jump back up following a different crop, especially a forage. But it isn't always true. In the western reaches of the corn growing region (eastern Texas to eastern North Dakota), corn yields better with nitrogen fertilizer than after a legume, since the forage uses up too much soil moisture.[1] Continuous corn has been grown for over 100 years on some test plots at the University of Illinois. On sections that received no extra treatments, the yields *are* pathetic, but on sections that received adequate amounts of lime, manure, and other fertilizers, yields are nearly as high (93%) as plots in a rotation.[2]

So there is no simple answer. Continuous row crops do tend to deteriorate the soil *unless* you are careful to keep building it up. A rotation can build soil structure and organic matter. If the market price of corn is low, it may be your best bet. But if you live in a moist climate (or can irrigate) and need corn to feed animals or to

[1] *Advances in Corn Production,* 1966, p. 14.
[2] *Modern Corn Production,* 2nd ed., 1975, p. 27.

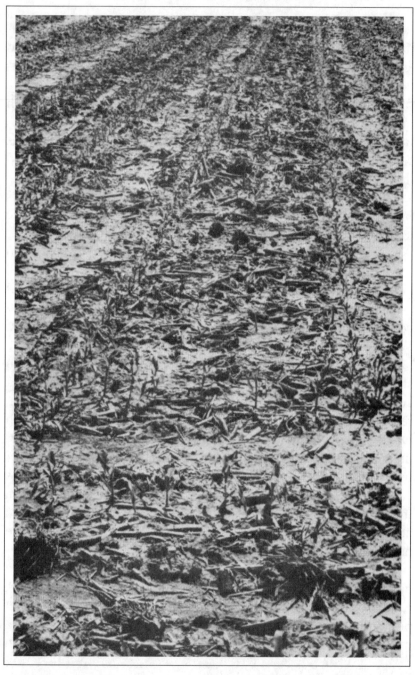

A no-till field.

sell, you can grow continuous corn successfully. You can even do that and "rotate" by growing an overwintering rye crop to build up organic matter. Soil fertility and soil conditions have to be closely monitored, and life-promoting methods of soil management must be used to produce top quality corn. The next chapter will show you how.

To till or not to till. Conventional tillage methods (plowing followed by disking and/or harrowing) served the American farmer pretty well for decades, at first powered by horse or mule, then by tractor. But some farmers were unwise. Steep slopes that should not have been planted to row crops eroded, and improperly fertilized soil became depleted. Also, the price of gasoline and hired men went sky high. Was this the fault of the plow and harrow?

Nevertheless . . . modern technology to the rescue! Ta-Daa . . . NO-TILL, the answer to all the soil's ills! Guaranteed to stop erosion in its tracks, build up soil organic matter, save tank cars of fuel, and produce yields that are "the same or higher" than with conventional tillage.

That's what is poured out for public consumption, but let's look at the rest of the story. Various reduced tillage systems, whether called no-till, conservation till, or minimum till, leave a mulch of crop residues on the surface (part may be disked in) to reduce erosion and conserve moisture; to plant (but requiring special planters) without ordinary tillage and seedbed preparation, and to reduce field traffic, saving fuel and reducing soil compaction.

Sounds great so far. But listen to what else is involved. To fight increased threats from weeds and pests, increased applications of herbicides and insecticides are required. Although they may not be killed, crops and the very important beneficial soil organisms are to some degree affected by these poisons.

Sorry to differ with the propaganda, but yields simply aren't always "the same or higher" than with conventional tillage. The mulch on top of the soil keeps soil one to eight degrees cooler than otherwise, plus toxic materials present in crop residues and produced by soil bacteria under wet conditions retard germination and stunt growth. The organic matter mostly stays near the surface, where much of its nutrients are lost by oxidation into the air. With little or no organic matter getting into lower soil layers, clay soils especially can become too "tight" underneath, with reduced water penetration. Surface-applied fertilizers do not encourage deep root systems, and nitrate release may be reduced, causing a nitrogen deficiency for the first two or three years. Weeds and insects are always a worry. Yields on fine textured, poorly drained soils and in northern climates are seldom as high as with conventional tillage.[3] Still sound so great?

Although reduced or no-tillage can be made to work in certain soil types and climates, the sad fact is that most of the desirable results of reduced tillage (erosion control, better soil structure or tilth, decreased compaction, increased water retention, reduced weeds and pests) can be obtained by other methods, *without* the drenchings of poisons. This book will show you how.

A step backward? Wanna hear a good one? Believe it or not, more and more farmers are abandoning hybrid corn and going back (or is it forward?) to open pollinated varieties, similar to the kinds their grandfathers probably grew—and the Indians who met the Mayflower!

Now why would a progressive farmer want to give up one of the most widely used marvels of modern agricultural science— hybrids? Aren't hybrids far superior to open pollinated varieties in yield, standability, and disease and pest resistance? Why return to the Dark Ages?

Before you split your sides laughing, let's take a closer look at hybrids and open pollinated varieties. First of all, what is a hybrid? A hybrid is the offspring that results from the crossing of two parents that are unlike in their hereditary characteristics. If you want to get picky, you could say that every person on the earth is a hybrid, since our parents differed in many characteristics, but what we usually mean in agriculture is a variety of plant (or animal) that is produced when two different *inbred lines* are crossed (or two different hybrid lines). Inbred lines are varieties that are

---

[3] Univ. of Wisconsin-Extension publications A3091, 1980, and A3001, 1979; *Bacteriological Reviews,* vol. 28, no. 2, 1964.

self-pollinated for several generations to produce a plant having certain desired traits or not having undesirable ones. Inbred lines are generally poor-producing, but when the right two inbreds are crossed (one is detasseled to eliminate the wrong kind of pollen), the resulting hybrid exhibits *hybrid vigor,* a dramatic increase in yield and other desirable traits, such as early maturation or disease resistance. Unfortunately, because of the way genes are inherited, planting seed from a hybrid in future years does not produce all offspring with the desirable hybrid traits, but a certain proportion of the offspring reverts back to the inbred characteristics. Thus, you have to buy new hybrid seed each year.

In contrast, open pollinated varieties "breed true," or do produce the same type of offspring year after year, allowing you to grow your own seed.

Open pollinated corn is generally believed to yield poorly and be more susceptible to lodging and diseases. Well, sometimes that's true, but not always. "With respect to yield, it should be emphasized that hybrid corn does not necessarily produce bigger or better ears than the best open pollinated plants."[4]

Hybrids are designed to be as uniform as possible— ears all the same height, maturing the same time—and to give consistent yields under certain soil and climatic conditions. But hybrids are limited. They have those particular genes (traits) that the corn breeders designed them to have, and they don't have as much individual variation as open pollinated corn, which allows some open pollinated plants to do well in almost any condition of soil and weather. But since open pollinated plants do have more variability, therefore in a stressful year, some plants won't do very well (while others will). Hybrids are "fine tuned" for certain specific soil and weather conditions, so if you pick the right hybrid for your soil AND if the weather is normal, you get terrific yields, but if something goes wrong—you have a total disaster—all plants are the same. The best example of this was in 1970, when 80% of the U.S. corn crop was susceptible to southern corn leaf blight and weather conditions favored the

[4] *Botany*, 4th ed., by C. L. Wilson & W. E. Loomis, 1967, p. 342.

wildfire spread of the disease, devastating the crop. That disaster happened because only a handful of corn hybrid varieties from only two races of corn are commonly used to produce the commercial hybrids. There are a few hundred major races of corn and about 50,000 varieties, a tremendous reservoir of genetic diversity.

Also, hybrids are usually designed with YIELD given first priority, then such things as lodging and disease resistance. Usually way down the list is quality, if it is considered at all—protein content and amino acid balance, vitamin and mineral content. Hybrid varieties are tested on different soil types and with different fertilizers to see under what conditions each variety does best. The trouble is, hybrids aren't designed for really good fertile, living soil; instead they do best under this or that nutrient imbalance or in wet soils or low organic matter. Hybrids do not have all the genes necessary to take up all the trace elements needed for healthy animal growth, and they often are out of balance in other nutrients. They easily develop excess nitrates under drought stress. Tests by Ernest M. Halbleib of McNabb, Illinois, revealed that hybrid corn did not absorb seven to nine trace minerals. None of the varieties tested took up cobalt, a mineral needed in vitamin B_{12}, which animals need for good health.[5] According to the University of Nebraska Experiment Station Bulletin 144 (1946), open pollinated corn makes chicks gain faster and have more protein in their meat than hybrids, and it makes far superior silage. Adolph Steinbroun of Fairbank, Iowa, had his open pollinated corn tested for nutrients. Compared to average midwestern test figures, his corn had 75% more crude protein, 875% more copper, 345% more iron, and 205% more manganese.[6]

More recent USDA tests for 13 major nutrients in various fruits and vegetables found that from 1950 to 1999, protein fell 6%, iron dropped 15%, vitamin C declined 20% and riboflavin fell 38%. Although not all crops tested may have been hybrids, some scientists believe the trend toward fast-growing, high-yielding crops is partly responsible for the nutritional decline, since fast-growing plants may not be able to absorb as many nutrients as slower-growing varieties.

[5] *Eco-Farm — An Acres U.S.A. Primer*, 1979, p. 38.
[6] Ibid.

The thing about open pollinated corn is that to do really well, it needs ideal soil conditions: fertile, well aerated, and the right moisture content. If your soil isn't in really good shape, open pollinated isn't for you—yet. But in the rest of this book, I will show you how to improve your soil so that you can start growing truly high quality corn.

I have seen beautiful open pollinated corn grown on a well balanced soil with less than 35 lbs/acre of added nitrogen. This was in northeastern Wisconsin (Manitowoc Co.), close to the northern limit for growing good ear corn. And it matured and dried down much earlier than the hybrids across the road.

Yields of open pollinated corn aren't always quite as high as hybrids—but the big difference is *quality*. We'll say more about that in Chapter 6.

Do you need GM corn? Since the 1980s, probably no new development has created more excitement—and controversy—than GMOs (genetically modified organisms). The first commercial GM (genetically modified) corn varieties, usually known as Bt corn, were released by a few companies in the mid-1990s, but soon one company, Monsanto, became the industry leader with its Roundup Ready corn. Bt crops contain a gene from the common soil bacterium *Bacillus thuringiensis* (thus Bt). These bacteria produce a toxin that when ingested by certain insect larvae, especially caterpillars, disrupts their digestive system, killing them by starvation. Organic growers and home gardeners have used Bt for many years as an effective non-chemical insecticide. The genetically engineered Bt corn is intended to mainly control the corn borer caterpillar and supposedly reduce the use of chemical insecticides. More recently, Monsanto has developed a GM corn that kills or resists the corn rootworm.

Roundup Ready crops have a gene that renders the plant resistant to the herbicide glyphosate (which used to be sold solely by Monsanto, but after their patent expired is now made by other companies). Thus a farmer can plant these crops and still control weeds. GM corn use has steadily increased in the U.S., with about 80% of the corn acreage now growing it.

GMOs are produced by using high-tech methods to insert one or more genes from one species (plant, animal or microbe) into another species, in order to produce a desired characteristic such as pest or herbicide resistance. The GM industry insists that their crops are totally safe (for food and in the environment) and that they are vital to help "feed the world."

Actually, they haven't lived up to the hype. Many experts worry that combining genes from two species can have unpredictable results, since the thousands of genes within one species' cells have developed a finely-tuned mechanism that regulates all aspects of growth and reproduction. There have been cases where GMO genes have escaped into the environment, causing ecological upset. An example is the wind-blown pollen of corn, with the Bt toxin killing monarch butterfly caterpillars. Also, some weeds have developed resistance to herbicides, including Roundup, sometimes called "super weeds."

An especially frightening side-effect of GM crops is that laboratory tests feeding them to animals (rodents) and livestock have found serious health problems, including crippled immune systems, pre-cancerous cell growth, liver damage, abnormal development of certain body organs, sterility, and premature death. Yet it is apparently OK for humans to eat such food, since already, without adequate testing, GM corn (and other crops) are consumed by nearly everyone on earth.

Because of these concerns, dozens of other countries either reject GM crops entirely or are very cautiously investigating them. This makes it very difficult for the U.S. to export farm commodities, resulting in millions of dollars of lost sales. Any contamination of non-GM grain shipments by GM grain is cause for rejection, which requires grain storage facilities to keep them separate. Also, yields of GM crops are often mediocre.

Surveys of GM-using farms have often found only very slight reductions in pesticide and/or herbicide use. Another serious problem is that the seed companies, which hold stringent patents on their GM products, charge high prices to farmers and do not allow them to save seed for future planting, with hefty lawsuits the result for transgressions.

Considerable experience by sustainable and organic farmers has shown that it is not necessary to grow GM crops to obtain high yields and to produce high quality, nutritious food. Healthy, vigorous plants have few pests, and weeds can often be controlled with little or no herbicide. GM corn—you don't need it!

CHAPTER 2

First Things First

So you want to grow top quality corn. Where do you begin?

Soil. The very most basic thing for growing really good crops is good soil. Soil that is not only high in fertility, but is alive with beneficial organisms. The ideal soil for growing corn is deep (six or more feet), medium-textured and loose, well-drained, high in water-holding capacity and organic matter, and able to supply all the nutrients the plant needs. Of course, not everyone has the perfect soil, and corn isn't so fussy that it can't do well on less than ideal soil. But I will show you how to build up your soil so that you can grow much better corn.

Climate. Corn does best with warm, sunny growing weather (75–86° F), well-distributed intermittent moderate rains, or irrigation (15 or more inches during the growing season), and 130 or more frost-free days. The U.S. corn belt has these soil and climatic conditions.

Humus. Even if the weather isn't ideal, a good, living soil with high humus content will often make the difference between a good crop and disaster, for humus allows soil to soak up considerable moisture and hold it for dry periods. It is often the case that one farmer who has been building his soil will have lush, green crops in a drought year, while his neighbor's crops have burned up.

Soil parts. An average, good soil should contain nearly one-half mineral particles, one-fourth water, one-fourth air, and a few percent organic matter. The minerals supply and hold some nutrients and give bulk to the soil. Water is necessary for plant growth and for the soil organisms, but not too much or too little. Air (oxygen) is needed by roots and beneficial soil organisms. Organic matter (humus and the living organisms that produce it) is a storehouse of certain nutrients, holds water, gives soil a loose crumbly texture, reduces erosion, buffers and detoxifies soil, and even helps protect plants from diseases and pests because of antibiotics and inhibitors produced by beneficial bacteria and fungi. Some of these friendly microbes also produce plant growth stimulators, others help feed nutrients directly to roots, and others trap (fix) nitrogen from the air—*free fertilizer.*

But things can go wrong. If the soil is short of air from waterlogging, low humus, compaction, or crusting, roots will suffocate or be "stunned," and the "bad guys," anaerobic bacteria, will take over and release nitrogen (denitrification) and produce several toxic substances, such as hydrogen sulfide, ammonia, aldehydes, and alcohols, when they decompose organic matter. **Tight and wet soils are one of today's worst enemies of good quality crops.**

Nutrients. To be healthy and produce excellent crops, a growing plant needs an adequate *and balanced* supply of over a dozen nutrients, mostly coming from the soil. Some are needed in larger amounts (the major nutrients: nitrogen, phosphorus, potassium, calcium), while others are needed in smaller amounts (the secondary and trace elements: magnesium, sulfur, iron, copper, zinc, manganese, boron, molybdenum, and chlorine). These plus carbon, hydrogen, and oxygen from the air and water are put together by the plant to form carbohydrates (sugars, starches, cellulose), fats, proteins, vitamins, and other miscellaneous products. Photosynthesis (powered by the sun's energy) and other metabolic processes accomplish these feats.

In a living, well aerated, fertile soil, the minerals, humus, and microorganisms should supply all of the plant's needs if there are no stresses from weather. But things often don't work out ideally, plus with the high populations of corn that are grown these days, some supplementation with outside fertilizers is usually necessary. Fertilizers will be covered in the next chapter.

Seedbed preparation. There are several basic methods and many variations of ways to till the soil and prepare a good seedbed for planting. An ideal seedbed is warm, moist, well aerated but firm soil. Old timers used to carefully plow and harrow the entire field to uniform fineness, but since only the row serves as the seedbed, some methods only prepare a seedbed in the row, leaving the soil between the rows rougher. This will provide a poor seedbed for weeds between the rows and allow the corn to get a head start.

The most commonly used tillage and planting methods are:

1. Conventional: plowing (in fall or spring) followed by disking, field cultivation, rotary hoeing, or harrowing. Requires more traffic; possible increased compaction. Fall plowing is best for fine-textured soils in the North; also it exposes hibernating insect pests to freezing weather. Do not fall plow on steep slopes subject to erosion. Crop residues should be mostly worked into the plow layer, and not deeply clean-plowed (over 8–10 inches). Moldboard

plowing where the soil is turned on edge is good for breaking sod and incorporating green manure crops. Disk plowing is good for dry regions and light soils; it may prepare a good seedbed in one pass. Chisel plowing is fast, good for well-drained soil, but leaves most trash on top (plant residues ideally should be worked into the upper several inches to one foot; otherwise they will not turn into humus). Chisel plowing is not good for sod and moist soils. Never plow too wet soil; serious compaction and loss of tilth results.

2. Listing: the lister ("middlebreaker") opens a furrow by throwing soil to the sides (by a double moldboard or disks); seed is planted in the bottom (later cultivations throw soil back into the furrow). It is good for medium-textured soils in dry and hot areas.

3. Ridge-planting: similar to listing except that seed is planted on the ridges. It is good for areas with abundant rainfall (ridges should follow the land's contour [across a hill, not up and down it] to catch rain and slow erosion). Also, ridges warm up fast in the spring if they run east-west.

4. Cultivator planting: after fall or spring plowing, final tillage and planting are combined in one operation. It is good for controlling early weeds if the soil has become crusted.

5. Reduced tillage, minimum tillage: there are several systems:

   a. Strip tillage: same as cultivator planting except that only the row is tilled (with rotary hoes, etc.). It is good for medium or coarse well-drained soils in the western corn belt: It does not work well if a living sod is left between rows.

   b. Wheel-track planting: spring plowing followed (within hours if possible) by planting, with tractor and planter wheels firming the seedbed. Can be done without plowing after small grains or row crops if the soil is moderately moist. It is good on light and moist soils, but crowds much field work into a short time.

   c. Plow-plant: plowing and planting are combined in one operation. It greatly reduces traffic, but requires good tilth or light soils.

d. Mulch tillage: leave crop residues on the surface, kill weeds with herbicide, till with sweep or chisel, then light disking, or use rotary tiller for entire operation; seed is planted with regular planter with disk openers or furrow openers on shoe-type planter. Conserves moisture, reduces runoff. It is used in dry areas, but is not good for fine-textured soils, which settle and become compacted.

e. No-till: used on grain stubble, sod, or a cover crop. Vegetation is killed with herbicide, planting is done with a furrow opener. Allows planting on wet soil and steep hillsides; decreases erosion (at least temporarily). It is not good for fine-textured, poorly drained soils and in the North.

Out of these tillage and planting methods, choose the one that suits your climate, soil type, and machinery. *You should think twice before using systems that require high herbicide use* (see also my comments in Chapters 1 and 4).

Seed and variety selection. First of all, you should always plant high quality seed with a high germination rate (above 98%). Seed quality depends partly on the health and vigor of the plant that produced it, as well as the care used in picking, drying, and sorting it. Seed grown under weather stress (drought, too wet, etc.) should be avoided.

It is very important to choose seed that is suited to your climate and soil, and for a desired maturation time. If you are growing corn for silage, choose a longer maturation time than if you want grain, since silage is harvested at an earlier stage of maturation. Another thing to remember is that 120 day corn, for example, may not mature in exactly 120 days. Growth can be slowed by unfavorable weather (too cold or too hot, drought, flooding) or nutrient deficiency, and it can be speeded up by optimal weather and soil nutrition, sometimes by as much as ten days to two weeks. A more exact way of measuring time to maturity is growing degree days (GDD). Growing degree days are calculated for each 24-hour day and added throughout the season. The formula to use is:

$$\frac{\text{maximum temp.} + \text{minimum temp.}}{2} \text{ minus } 50 = \text{GDD}$$

If you don't have a minimum-maximum thermometer, you can use figures from the nearest weather station. If the temperature goes below 50°F, substitute 50° for the minimum temperature, and if it goes above 86°F, substitute 86° for the maximum.

Seed should be chosen that contains desirable nutrient characteristics if you are growing grain or silage to feed animals. Among the hybrid varieties, there is a wide difference in protein quality and quantity. Most corn varieties are low in three essential amino acids, lysine, methionine, and tryptophan, so new hybrid varieties, generally called high-lysine corn, were developed. But they have their own problems: low germination and yield, and high moisture and kernel damage. Plus no hybrid contains as good a balance of trace minerals as open pollinated corn. Other special hybrids include (1) waxy maize, which contains high amounts of amylopectin starch and is said to produce more efficient gain in beef cattle; (2) high sugar corn (sweet-stalk corn), which is said to be good for silage, but it is lower yielding; (3) high oil corn, which is said to be good for fattening hogs, although it produces a softer backfat; and (4) upright-leaved corn, which uses light better when planted in close rows.

Many farmers are switching to open pollinated varieties. New methods for improving them are being developed. Different varieties have different characteristics; for example, white corn is high in carbohydrates, yellow is high in vitamin A, and colored ("Indian corn") is high in minerals.

In general, higher test weight seed is of higher quality. To go through a planter well, seed should be graded for uniform size. Seed treated with fungicide and insecticide is good insurance, especially if your soil isn't in ideal condition. Sometimes adding ordinary sugar to the planter box will take the place of fungicide treatment, plus stimulate soil bacteria.

Population and row width. There are many factors that govern optimal population and row width, ignoring the obvious one of the type of machinery you own. Very important are moisture and nutrients, for high population corn needs high fertility soil and adequate moisture. In general, with average rainfall, you can plant higher populations in the northern and eastern U.S. than in the southern and western areas of the corn belt, unless you can irrigate. If moisture and nutrients are adequate, narrower rows and higher populations give higher yields. Drilling, with accurate within-row spacing, gives better yields than hill-dropping or check-planting. Higher populations can be planted if corn is grown for silage or fodder than for grain. Average populations are about 17–18,000 per acre, low populations are about 12,000 per acre, and high populations can run up to 25–30,000 per acre. Typically, only about 85% of planted kernels reach maturity, so here is a table giving kernels per acre to plant and kernel spacing for drilled corn (adapted from R. J. Delorit, L. J. Greub, & H. L. Ahlgren, *Crop Production,* 4th ed., 1974, p. 105):

Plant Population Per Acre	Kernels/A at 85% Maturing	Kernel Spacing					
		20" row	28" row	30" row	34" row	38" row	42" row
12,000	14,100	22.2"	15.9"	14.8"	13.1"	11.7"	10.6"
14,000	16,500	19.0	13.6	12.7	11.2	10.0	9.1
16,000	18,800	16.6	11.9	11.1	9.8	8.8	7.9
17,000	20,000	15.6	11.2	10.5	9.2	8.3	7.4
18,000	21,200	14.8	10.6	9.9	8.7	7.8	7.0
19,000	22,400	14.0	10.0	9.3	8.2	7.4	6.7
20,000	23,500	13.3	9.5	8.9	7.9	7.0	6.4
21,000	24,700	12.7	9.1	8.5	7.5	6.7	6.0
22,000	25,900	12.1	8.7	8.1	7.1	6.4	5.8
23,000	27,100	11.6	8.3	7.7	6.8	6.1	5.5
24,000	28,200	11.1	8.0	7.4	6.5	5.8	5.3
25,000	29,400	10.6	7.6	7.1	6.3	5.6	5.1
26,000	30,600	10.2	7.3	6.8	6.1	5.4	4.9
27,000	31,800	9.8	7.1	6.5	5.9	5.2	4.7
28,000	33,000	9.4	6.9	6.3	5.7	5.0	4.6
29,000	34,100	9.1	6.7	6.1	5.5	4.8	4.4
30,000	35,300	8.8	6.5	5.9	5.3	4.7	4.3

Although higher populations give higher yields under ideal conditions, if there are stresses from adverse weather or if the soil runs out of readily available nutrients, yield and test weight will be reduced, as will ear size, leaf area, number of ears, and protein and oil content. Silking will be delayed, leading to poor pollination. Lodging, stalk rot, corn borer, and other problems will be increased. Nutrient deficiency throughout the growing season can often be corrected by side dressing and/or foliar feeding (see Chapter 3). But considering the unpredictability of the weather, it may be best to avoid the temptation to plant very high populations and risk getting a poor crop.

Planting. In general, earlier planting produces better yields, but that can be carried too far. Everyone wants to be "first on their block" to get their corn planted, and often the soil is simply too

How to Estimate
Plant Population Per Acre

An accurate estimate of plant population per acre can be obtained by counting the number of stalks in a length of row equal to 1/1000 of an acre. Make at least three counts in separate parts of the cornfield, figure the average of these samples, then multiply this number times one thousand.

Row Width	Row Length Equal to 1/1000 Acre
20"	26' 1"
24"	21' 10"
28"	18' 8"
30"	17' 5"
32"	16' 4"
36"	14' 6"
38"	13' 9"
40"	13' 1"

cold for germination, or a late cold spell injures seedlings or slows growth. Corn will hardly germinate below 50°F soil temperature, and the best temperature is about 68°. Tests have shown that at 50–55°, corn takes 18–20 days to emerge; at 60–65° it takes 8–10 days; while at 70° it only takes 5–6 days. Best seedling growth occurs at 86°F.[1] It is best to check the soil temperature at a two-inch depth to be sure the soil is warm enough. Checking in the morning will give a truer reading, since the soil temperature at that depth

[1] *Corn and Corn Improvement*, 1979, p. 602.

Number and Length of Rows in an Acre

One way to estimate the number of acres in a cornfield, or portion of a cornfield, is by computing the length of the rows and the distance between rows. The following table shows the number and length of rows in one acre:

Length of Rows in Rods*	If distance between rows is:						
	20"	24"	30"	32"	36"	38"	40"
40	39.6	33.0	26.4	24.7	22.0	20.8	19.8
60	26.4	22.0	17.6	16.5	14.7	13.9	13.2
80	19.5	16.5	13.2	12.7	11.0	10.4	9.9
100	15.8	13.2	10.5	9.9	8.8	8.3	7.9
120	13.2	11.0	8.7	8.2	7.3	6.9	6.5
140	11.3	9.4	7.5	7.0	6.3	5.9	5.6
160	9.8	8.2	6.6	6.2	5.5	5.2	4.9

*One rod equals 16.5 ft.; 40 rods equals 660 ft.

may rise 5–15 degrees on a sunny day. When soil temperatures reach about 55–59° for several days—plant. Planting should not be delayed if the soil is warm enough, so that the corn can make its vegetative growth before hot, dry weather, which can interfere with silking and tasseling. However, later planted corn, especially open pollinated varieties, can often catch up with early planted hybrids if soil fertility is high enough.

Seed should be planted deeper (2–4 inches) in light or lumpy soil, and shallower (1$\frac{1}{2}$–3 inches) in heavy soils or a good seedbed. Proper moisture at planting depth is most important, then comes temperature. In dry soils, you may have to plant as deep as 3–4 inches in clay or 5 inches in sand to have enough moisture. For very early planting, plant $1/_2$ to 1 inch shallower than normal (if moisture is adequate) to avoid cold temperature.

CHAPTER 3

Fertilizing

Corn, much more than many other crops, needs high soil fertility, especially at modern high populations. But there are several factors to consider.

Before about 1950, most corn fertilizer came from manure and legumes, with considerable lime being used. Corn crops were in rotation with other crops, so that only $^1/_3$ to $^1/_2$ of the cropland was planted to corn at a time. Since then, there has been the overwhelming tendency to grow higher yields and continuous corn, and to use increasing amounts of synthetic commercial fertilizers and pesticides. These trends have been accompanied by decreasing humus content, harder "deader" soil, and worse erosion, weed, pest, and disease problems. Could there be a connection?

Absolutely! Wrong methods of fertilization and soil management can make the difference between healthy, fertile, "living" soil and "dead," sterile soil. The two big factors are adequate humus and a high population of beneficial soil organisms (stop and review what we said about them in Chapter 2). The beneficial organisms can't live and do their jobs if the soil is saturated with toxic substances. Toxic substances not only come from pesticides (herbicides and insecticides), but also are produced by anaerobic bacteria in tight or waterlogged soil, and *from certain harmful fertilizers*. Yes, that's right, there are good and bad fertilizer materials, and right and wrong ways of using even the good ones.

Problem causers. Some of the problem causers include (1) anhydrous ammonia, which is toxic to nearby roots and soil organisms, causes large pH changes (first high, then low), and destroys humus (by making it soluble and leachable), causing hard soil; (2) urea and diammonium phosphate (DAP), which can release ammonia when they break down, causing some of the same problems as ammonia; (3) chloride-containing fertilizers, especially the commonly used potassium chloride (muriate of potash), which is a high-salt fertilizer and therefore can cause root damage and which kills or inhibits beneficial soil bacteria; (4) high magnesium materials on soil with adequate magnesium, primarily dolomitic limestone, which is slow to become available, and can lead to hardened soil and can release soil nitrogen reserves; and (5) wrong use of animal manures, either over-application or plowing or injecting them too deep, causing toxin production by anaerobic bacteria.

Wet or dry? Then there is the question of liquid vs. dry fertilizers. Which is best? While it is true that liquids offer some advantages, namely that they can be handled easily and applied rapidly, and can easily be custom mixed, still they have some disadvantages and are not as versatile as dry fertilizers, which can also be custom blended. For one thing, liquids cannot be mixed in very high analysis solutions in blends (generally less than about 30 units of fertilizer) because of crystallization, or "salting out," of the components (although slurry suspensions are possible). So you are hauling and spraying a lot of water. Another more serious problem is the possibility that dry conditions could follow application and result in the fertilizer salts being concentrated and "burning" the seedlings. Liquids should never be applied on the seeds, for that reason. On the other hand, dry fertilizers will not be activated in dry weather, eliminating possible damage to plants (although dry fertilizers can also damage seedlings if placed too close to the seed). Dry fertilizers can become available to plants just as rapidly as liquids when there is normal soil moisture, since dry fertilizer particles attract moisture and quickly dissolve.

Liquids can be used effectively if care is used in the choice of materials being applied and of application methods. Chloride-containing materials should always be avoided. You should have a sample of any liquid fertilizer tested to be certain it is chloride free (or at least below 2% chloride).

Fertilizer needs. Does it make much sense to apply a heavy dose of starter fertilizer just before planting? Stop and think, when does the growing plant need the most nutrients, when it is a seedling, or when it is actively growing and maturing? Most of the fertilizer in starters is not needed early in the season, and much of it (mostly nitrate nitrogen) can be lost (from leaching and denitrification) before the plants have a chance to use it. Also, too much nitrogen early in the season promotes lodging and produces a poorer balance of amino acids in the grain.

According to studies of nutrient uptake by corn, it can be seen that the plant's need for soil nutrients begins at near zero (during germination) and gradually increases throughout the growing season. Therefore, the optimal way to supply nutrients to the crop is to either use slow-release fertilizers (ammonium nitrate or organic matter) or to use split applications throughout the season (by side dressing and/or foliar feeding). Now, if the soil is healthy and "alive" with beneficial organisms and has plenty of humus and no toxins, a proper balance of nutrients will be made available to the plants in increasing amounts through the growing season. This is because the activities of the soil organisms increase with increasing temperature. Considerable nitrogen will be taken from the air by non-symbiotic nitrogen fixing bacteria and algae. Soil microbes will even help feed nutrients to the roots, and some will protect roots from diseases and pests. That's the way it *should* work. With less than perfect soil conditions and with high plant populations, supplementary fertilizers will be needed, but they should be the types that are compatible with soil life, not toxic to it.

Requirements. Let's see briefly what elements a growing corn plant needs and their functions in the plant. The elements needed in largest amounts are called major elements. Corn needs considerable nitrogen (N is its chemical symbol), mainly in the nitrate form, although an application of an ammonium-containing fertilizer about 45 days after planting will more effectively help the plants "switch over" to flowering and grain production activities. Nitrogen is needed by the plant for certain enzyme functions and

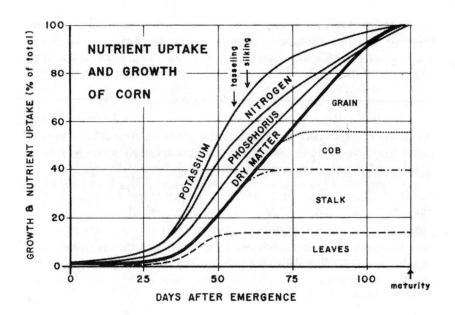

for protein production, and it is a necessary part of chlorophyll, nucleic acids, vitamins, and several other molecules.

Phosphorus (P) is a very important element for general growth, photosynthesis, and metabolism. It carries energy from one part of the cell to another and supplies energy for the transport of nutrients and food from one part of the plant to another. It is a part of cell membranes, nucleic acids, etc. It is necessary to grow high quality crops and is especially needed by young and actively growing plants.

Potassium (K) is used in enzyme reactions, protein and chlorophyll production, food transport from one part of the plant to another, and in regulating water balance in the plant's tissues. Potassium improves disease resistance if not too much is applied.

Calcium (Ca) is vitally important for cell division and root hair growth, enzyme activities, and for normal cell walls. Adequate calcium improves plant resistance to diseases and gives better quality crops with regard to animal (or human) nutrition.

Secondary elements are needed in lesser amounts than the major elements. Magnesium (Mg) is needed as part of chlorophyll and for the structure of nucleic acids, cell membranes, and cell structures that produce proteins. It is also required for some enzyme functions.

Sulfur (S) is a necessary part of certain amino acids (methionine, cysteine, cystine) and is important in many enzyme functions. (Elemental sulfur should not be applied to the soil; sulfur should be supplied in the sulfate form.)

The remaining elements are needed in very small amounts and thus are called trace elements or micronutrients. Each one has different functions, but in general most of them are involved in enzyme activities in metabolism and photosynthesis. The trace elements include iron (Fe), zinc (Zn), copper (Cu), boron (B), manganese (Mn), molybdenum (Mo), cobalt (Co), and chlorine (Cl).

Balance. It is very important for crop health and high quality yields that nutrients be available to the plants *in the proper amounts and in proper balance.* Too much or too little of some elements can cause deficiencies of others, as can too high or low soil pH. For example, adding potassium may cause a magnesium deficiency, while high pH can cause zinc, manganese, iron, and boron deficiencies by tying up these trace elements on soil particles.

Added to this is the problem that some experts tend to recommend amounts of some fertilizers that lead to a balance of elements that produces good yields (quantity), *but poor quality.*

Testing. To find out what fertilizers need to be added to your soil, it is important to have your soil tested. But the trouble with soil testing is that there are dozens of ways of doing it and interpreting the results, and some of them are very misleading. Most testing labs use methods that test the soil under artificial (dried, finely ground) rather than field conditions, and that determine the total amounts of some nutrients rather than what is readily available to the plant roots at any one time. Also, many labs recommend too high amounts of some fertilizers, just to "be on the safe side."

The most realistic method of soil testing uses weak extracting fluids that simulate the nutrient extracting power of roots. This soluble testing shows you what soil nutrients your crops could absorb at that time. By testing more than once during the year, you can get a better idea of what is happening in your soil than by testing once every year or two. Unfortunately, few soil testing labs routinely use these methods, although most will run soluble tests if requested.

Pre-plant fertilization. If soil tests reveal that your soil needs large amounts of certain fertilizers and you want to (or your budget allows you to) go on a soil building program, then the proper fertilizers should be broadcast. On the other hand, in-row pre-plant fertilizer and split applications of side dressing during the growing season are a very economical way to grow a good crop one year at a time.

Such soil building additives as crop residues, animal and green manures, compost, lime, and soft rock phosphate are best added in the fall, although the latter three can be applied in the spring. Manures and compost should always be plowed or disked down several inches, but not deeper (they need to be in the aerobic, or aerated zone). Up to 6–10 tons per acre of fresh animal manures are beneficial for corn, applied in the fall (only $\frac{1}{2}$ to $\frac{1}{3}$ as much poultry manure should be applied as cattle manure). Ammonium sulfate is a good fall fertilizer because it helps warm up the soil faster in the spring and keep it cool in hot weather.

Especially in cool, wet soils (in the North, and for early planting), an in-row starter fertilizer is valuable. It gets a uniform stand off to a quicker start, helping the corn to stay ahead of weeds and hastening maturity. The best placement for a band of starter fertilizer is about $1\frac{1}{2}$ to 2 inches below and to one side of the seed. That way the early roots can encounter it easily. "Pop-up" fertilizers (low levels of fertilizer on the seed) run the risk of fertilizer "burn" in dry weather.

Adequate levels of nitrogen and phosphorus are especially important early in the season. Since there may not be high release of these elements by soil organisms at cool temperatures, good synthetic sources are ammonium sulfate, ammonium nitrate, and mono-ammonium phosphate, depending on what the soil tests in-

Fertilizers can make a lot of difference to the all-important root system.

dicate is best. Soft rock phosphate, a natural product, also supplies phosphate in a form readily used by plants.

Mid-season fertilization. OK, so let's say you prepared a good seedbed, applied your starter, and planted your seed. You now have a good looking stand of young corn plants "off and running." Can you now take a vacation and come back at harvest time? Not if you want to be farming five years from now.

A growing crop should be monitored closely to see how it is doing and to check for developing problems. Large reductions in yield and quality can occur from things that happen early in the life of the plant. In corn, the period of tasseling and silking is especially critical. Don't forget the pollen and cob development are first begun way back when the plant is only knee high (review Chapter 1).

Therefore, no matter how busy you are, it is worthwhile to walk around in your fields at least weekly. Learn to be alert for signs and symptoms of problems. Don't just look at the leaves and stalk; also check the other half of the plant—the root system—and the soil it is growing in. Carry a shovel and dig up a plant, shake the dirt off the roots and see if you have a healthy set of roots— main roots, feeder roots, and root hairs. Are there brown or dying roots, or roots whose outer layer stripped off when the plant was pulled from the ground? This may indicate possible toxic or anaerobic soil, or salt damage. If the root system isn't healthy, the rest of the plant won't be either, and it can't function effectively to make food and produce good grain. Sick plants are easy prey for pests and diseases. More about this in Chapter 4.

Since the greatest need for soil nutrients comes in the latter $^2/_3$ of the corn plant's life, a mid-season soil test should be run to see if adequate nutrients are available, and in the right balance. As we mentioned earlier, soluble testing can give this information. If soil fertility isn't what it should be, mid-season fertilization can make the difference between a near disaster and a terrific yield. What happens in all too many soils (those without adequate life, humus, and air) in the last part of the growing season is that the soil just runs out of adequate available nutrients. The cob doesn't fill completely, or the kernels dent excessively, or the grain won't dry down naturally. Weather stresses (heat, drought, excess rain, etc.) make the problems worse. Generally, pests and diseases take a large toll.

There are two ways of supplying an extra boost, or a "spoon feeding" for a crop while it is growing: side dressing and foliar feeding. They have different purposes and capabilities. Side dressing, either with liquid or dry fertilizers (dry is preferred, see earlier discussion in this chapter), is able to supply moderate amounts of major, secondary, or trace elements, as well as to give the plants a "shot in the arm" of a high energy fertilizer, which will stimulate them to take up more nutrients from the soil (which means the soil must have enough nutrients available or else growth will be slowed). For example, in a Door Co., Wisconsin, corn field which had sufficient calcium, phosphorus, and other nutrients, I have seen beautiful corn with well-filled ears (two per stalk) in mid-August after side dressing with 0-20-0 (superphosphate). Foliar sprays are mainly used to supply secondary or trace elements, but can also be used to "energize" plants.

Side dressing should be done starting when the corn is about knee high, or about 40–45 days after planting, and can be cost-effective even when done two or three or even four times during the middle of the growing season. It can even be done after tasseling if you have a highboy tractor. A convenient way of side dressing is to do it during cultivation by mounting fertilizer boxes on your cultivator. In side dressing, a small amount of fertilizer (perhaps 100 to 200 lbs. per acre) is applied between the rows on the surface (fertilizer should not be allowed to get into the whorl of the corn plant). Even a dry fertilizer on the surface will dissolve (from rain or dew) and become available to roots just below the surface. Side dressing is a very valuable aid to growing top quality corn if done properly. It can reduce wasted fertilizer that would leach away from a too-heavy pre-plant application, supply nitrogen during the time it is most needed, and give plants an extra boost of "energy" when it is vitally needed. However, the soil still should have an adequate supply of the important elements of phosphorus, calcium, and potassium before side dressing is done, for side dressing is best done to activate plants to absorb more nutrients that should already be available in the soil. An application of side dressing containing ammonium nitrogen at 40–45 days after planting will aid the plant to "switch over" to producing tassels and cobs.

On the other hand, foliar feeding is a little trickier to do effectively, and may not be cost effective on a crop such as corn. Ordinary field sprayers do not produce a very fine spray, and thus are wasteful of material. More expensive sprayers atomize or homogenize the spray, so you can sometimes spray a field for less than a dollar per acre of materials, especially if you mix your own spray formula. But that is too complex a subject to cover in this book. However, a good "all-purpose" spray is a mixture of about $^3/_4$ liquid fish and $^1/_4$ seaweed, diluted 2 gal. in 100 gal. water (deionized or soft water is best). The fish and seaweed mixture should be acid (pH 5-6.5); first adding 1 pint to 2 quarts of liquid phosphoric acid to the 100 gal. of water will take care of that. The liquid fish should be strained to keep from plugging your sprayer.

The trace element most often deficient in corn is zinc, especially in cold, wet soils and at high pH. You should never add trace elements just because you think they *might* be deficient, since they are needed in only very small amounts, and too much can be toxic. A soil test should be done and only the recommended amount should be applied. "Shotgun" mixtures of trace elements can be detrimental. A list of trace element leaf deficiency symptoms is found in Chapter 4.

If you want to try foliar spraying, a good way to see if the spray will help your crop is to spray some on several plants with a hand sprayer or squeeze bottle. After a half hour, test the sugar content of the sprayed plants with a refractometer (see Chapter 6) compared to unsprayed plants. If sugar increases, spray. The most effective time to spray is early in the morning (3–4 a.m. is even better!), since that is when plants take in the most materials through their leaves. Be sure that your sprayer has had no herbicides or insecticides in it.

CHAPTER 4

Problems

Weeds. *Corn borers.* **Stalk rot.** *Moldy corn.* You've all seen these and other problems that can (and usually do) afflict the corn grower. Most farmers think that insects and weeds are *inevitable,* so they spray their fields with toxic pesticides ahead of time, "because the bugs and weeds are going to come anyway."

But are these problems really inevitable? What is their *cause?* If we can identify and eliminate their cause, the problems will cease to exist! It's that simple.

The experts tell us that diseases are "caused" by pathogens—certain bacteria, fungi, viruses, etc., and that insects and weeds "cause" crop damage and yield losses. In a limited sense they are right, *but why are the bugs and weeds there, and why do the pathogens attack?* Not all fields and not all individual plants have these problems. So the problems are *not* inevitable.

*The ultimate **cause** of most crop problems is "sick" soil.* There is abundant evidence that soil fertility and soil conditions play a major role in controlling the health and yield of plants. Everything from rapid growth rate, to disease and pest resistance, to vitamin and protein content, to mold-free storage can be correlated with "healthy" soil. Even the effects of adverse weather (drought, excess rain, wind) can be greatly reduced by having your soil in good condition—loose and crumbly, with adequate humus and abundant beneficial organisms.

Insects, diseases, and molds are nature's "clean-up crew," there to eliminate the sick and unfit. They are the symptoms, as are weeds, of unhealthy soil and other stresses, such as adverse

weather. Let's look at some of the problems that afflict the growing crop and how they can be controlled or prevented (some other
problems will be covered in Chapter 5).

Weeds. Weeds are generally the first battle of the growing season. They seem to be especially bad in row crops such as corn.
There are more kinds of noxious weeds today than there were 50
years ago, and thanks to herbicides, we now have many weed varieties resistant to the most powerful poisons we can throw at them,
and even populations of soil bacteria in some areas of the corn belt
that gobble up certain herbicides before they can kill the weeds!

The simple truth is that *nearly all weeds grow best in poor
soil—unbalanced, tight, depleted, "dead" soil.* And that is just
what our modern agricultural methods have created over the last
few decades. In loose, fertile soil with the right balance of nutrients
and beneficial organisms, weeds just don't grow well, believe it or
not. And crops do grow well, and can out-compete the weeds.

So how can you keep weeds under control? There are several
ways. First of all, tillage methods that prepare a good seedbed in
the row but leave a rough, poor seedbed between the rows (plus
the use of a starter fertilizer) will allow the corn to sprout first and
get ahead of the weeds. If weeds get a head start, many species
release chemical secretions (phytotoxins) that inhibit crop growth,
cutting yields (the process of one plant inhibiting another is called
allelopathy). Many crops can do the same to weeds if they sprout
first. Weeds also reduce yield by competing for light, soil moisture, and nutrients. In Minnesota, grain yields have been reduced
from 16 to 93% (51% average) by weeds.[1] Weeds grow faster than
corn in cool weather, so early-planted corn is at a disadvantage.

A second weapon in the fight against weeds is timely cultivation, with shovel, rotary, or disk cultivators, rotary hoes, spike
tooth and spring tooth harrows, etc. Some weeds are easiest to
control when small, when their roots are shallow and the corn
roots are already deep. Rotary hoes and rotary cultivators are most
effective when the corn is small, but after it is over six inches tall,

* Lamb's quarters and redroot pigweed are among the few weeds that indicate good soil.
[1] *Corn and Corn Improvement,* 1976, p. 657.

shovel cultivators are most effective. Be sure to only cultivate deep enough to kill weeds and not to prune the corn roots. As the season progresses, stay toward the middle of the row. Cultivation also has the benefit of breaking a crust and aerating the soil—which can greatly stimulate crop growth and health. If your soil has a crust, it is wise to cultivate even if there are few weeds to kill. A "dust mulch" formed by breaking a crust is also a good weed inhibitor. Another excellent tool to use is a disk hiller (pairs of disks, one on each side of the row with cultivator sweeps between rows), which will throw soil into a ridge against the stalks of older corn plants. This process will help control weeds, stimulate corn prop root growth, and form a furrow between the rows that will trap rainfall (hilling and row direction are therefore best at right angles to the slope of a hill). Other unusual weed cultivation methods are burning them to death with propane-fueled flame cultivators (good for small in-the-row weeds) and electrocuting them with a 15,000–20,000 volt electrostatic rig (which works on weeds taller than the crop, but the equipment is expensive).

One seldom-used method of cropping corn can reduce weeds; that is, intercropping, planting a different crop between the corn rows, such as soybeans or other legumes, forages, or sod. Generally, corn yields are not reduced (although wider rows must be used), but the other crop will suffer from shading (but if a forage or grass will be in rotation the next year, then it will already be established). Experiments with planting alternating narrow strips (6 or 12 rows) of corn and soybeans have shown that corn yields can be increased 15 to 50 bushels per acre because of increased sunlight available to the outside rows of corn. The soybean yield is reduced 10 to 20%, however.[2]

The use of herbicides to control weeds is at best a crutch, which only creates more problems than it solves. Besides killing plants, herbicides kill or at least upset beneficial soil organisms, leading to reduced humus production and tight, anaerobic, toxic soil—an ideal soil for weeds. Some herbicides actually stimulate germination of weeds other than the ones they kill. Also, herbicides or their break-

[2] *Farm Journal*, 1983, June/July, p. 34-35.

down products can pollute ground water and get into crops—the food we and our animals eat— with often unknown consequences. If they absolutely have to be used to save a crop, herbicides can sometimes be reduced in dosage, especially if used with a surfactant (wetting agent). But with good soil, they may not be needed at all.

So that brings us to our last method of combating weeds and the ultimate solution in the long run—healthy soil. By having a loose, well aerated soil (plenty of humus is the most effective way), and the right balance of nutrients (high calcium; more available phosphorus than potassium), most weeds just won't want to grow. They can actually become sick and die! I have seen it happen, in less than one year of soil improvement.

Insects. There are dozens of kinds of insects that attack all parts of the corn plant—roots, stalk, leaves, and cob. As we have said, they are just part of nature's clean-up crew, eating the sick plants. Healthy plants will have little or no damage and even produce special chemicals that repel at least some pests. One such natural repellant is called DIMBOA, which is formed when corn tissue is injured and which repels young European corn borers, as well as the fungi that cause northern corn leaf blight infection and gibberella stalk and ear rots.

Insect pests also have their own natural enemies: other insects, parasites, and birds that keep them under control—if they aren't poisoned themselves by over-use of insecticides. If insecticides have to be used to save a crop, effective kill can often still be obtained by reducing the dosage. But many pests can be adequately controlled by non-toxic methods, many of which were used decades ago (see box on next page).

But the same as for weeds, the best long-range pest control is to have good soil which produces healthy, naturally resistant plants. Also, some hybrid varieties are more resistant to certain pests than others, and of course, some genetically engineered varieties contain the Bt gene and are resistant to corn borers.

Diseases. Although some corn diseases, such as stalk rot, are very widespread and common, in most years diseases only cause small yield reductions (7 to 17%; the 1970 corn blight disaster is

Non-Toxic Control
of Common Corn Insects

(from: *Advances in Corn Production,* 1966; and *Modern Corn Production,* 2nd ed., 1975)

1. *Control of insects attacking seeds* (seed corn maggot, wireworms, and seed corn beetles): delay planting until conditions are good for rapid germination (wheel-track plant in dry years); replant between old rows if great damage (then cultivate out the worst stand); fall plowing; drain low spots.

2. *Control of root-eating insects* (corn rootworm larvae, corn root aphids, white grubs, and grape colaspis larva): rotate or plant between old rows to escape large rootworm population; early plowing, disking, delayed planting, and extra phosphorus fertilizer for grape colaspis; disking to kill weeds and ant colonies for aphids; fall plowing and milky spore disease inoculant for white grubs.

3. *Control of insects feeding on underground part of stalk* (wireworms, billbugs, sod webworms, and cutworms): plant other crop than corn if breaking sod; early fall plowing; reseed if necessary, then cultivate out worst stand; drain wet areas; early spring plowing and planting for cutworms.

4. *Control of insects feeding on corn leaves* (armyworms, wooly bears, corn leaf aphid, corn flea beetle, grasshoppers, corn blotch leaf miner, chinch bug, cereal leaf beetle, 2-spotted mite, and thrips): early planting and resistant varieties for flea beetle; fall plowing and natural enemies for grasshoppers; don't plant beside small grains for cereal leaf beetle; natural parasites and fungus disease for aphids; control of thrips, mites, wooly bears, and corn blotch leaf miner seldom necessary; no good control for armyworms (but they will avoid a healthy crop).

5. *Control of insects feeding in whorl, stalk, and ear* (European corn borer, corn earworm, corn rootworm adults, stalk borers, fall armyworm, corn sap beetle, and picnic beetle): plant resistant varieties, midseason planting (not early or late), fall plowing, plow under or pulverize dead stalks, and bacterial disease inocu-

lant (Bt) for European corn borer and stalk borers; plant resistant varieties for corn earworm; control other insects and diseases
for sap and picnic beetles (they only feed on injured, unhealthy
plants); no good control for fall armyworm (but they will avoid a
healthy crop).

an exception). But they still do cause millions of dollars of losses
that could be avoided. Most infectious diseases of corn involve
fungi, with some involving bacteria and a few involving viruses or
other pathogens. They include seed rots, seedling blights, root and
stalk rots, smuts, leaf blights, bacterial wilt, leaf spot, corn rusts,
brown spot, ear rots, and several viral diseases.

The textbooks say that diseases break out when weather factors are just right (temperature, rainfall, humidity), when there are
soil drainage or pH problems, and *soil fertility* problems. Some
even say that *some diseases only flare up when plants are under
stress.* Those authors are right. All of those factors—weather,
wrong pH, waterlogged soil, and out-of-balance soil fertility put
plants under stress and are the real causes of disease. The "germs"
(pathogens) just come along to eliminate the sick plant. They cause
some symptoms in the plant (rotting stalk, leaf spots, etc.), so the
"experts" tell you to use a resistant variety or to spray a poison to
kill the fungus or bacteria rather than attacking the real causes of
the problem. *Good healthy soil is the key to growing healthy crops
that are naturally resistant to diseases.*

In the previous section, we mentioned DIMBOA, the chemical
that is released when healthy corn plants are attacked by spores
of northern corn leaf blight and gibberella stalk and ear rot, and
which inhibits fungal growth.

Another example of natural resistance involves root and stalk
rots, which develop in plants that have a low sugar content (low
rate of photosynthesis) and are under stress from drought, high
stand populations, leaf damage, cloudy weather, and fertilizer im-

balance.[3] The unhealthy plants can no longer resist fungal invasion of their roots, which spreads to the stalk. Various experiments have shown that corn will resist stalk rots when phosphorus and potassium are adequate and nitrogen not too high;[4] however, too much chloride (applied as potassium chloride) increased stalk rot and lodging after fungi had invaded.[5] In fact, one study found that only 60–120 lbs. per acre of potassium chloride, row applied, depressed corn growth, maturity, and yield.[6]

Seed rots and seedling blights are decreased when mature, high quality seed is planted and are increased from the stress of cold temperatures[7] and wet soil.[8]

It has recently been discovered that some soils naturally suppress certain plant diseases, giving much greater crop yields. Investigation revealed that in such soils, certain species of beneficial bacteria colonize plant roots and inhibit certain harmful bacteria that attack roots and allow diseases to begin. Furthermore, treating seeds with the beneficial bacteria increases plant growth and yield.[9]

So, can you see the importance of good healthy soil? Soil that is loose, well aerated, with abundant humus and beneficial organisms, and balanced fertility will produce healthy plants that naturally resist diseases.

Soil conditions. As we have said, most crop problems can be traced back to the soil. We have already covered the balance of major nutrients and beneficial soil organisms. Now let's mention some other soil problems that cause crop problems.

Crusting, cloddiness, tight soil, poor drainage, and waterlogging are all symptoms of poor physical structure and low humus content. Sometimes the problem is worst beneath the plow layer, where dense subsoil or a hardpan may be causing poor drainage

[3] J. L. Dodd, *Phytopathology,* 1980, p. 534–535.

[4] *Phytopathology,* 1960, p. 212–214.

[5] *Agronomy Journal,* 1967, p. 499–502; 1976, p. 425–426.

[6] *Ibid.,* 1958, p. 423–426.

[7] *Corn and Corn Improvement,* 1976, p. 397.

[8] *Modern Corn Production,* 2nd ed., 1975, p. 247.

[9] *Science,* 1982, vol. 216, p. 1376–1381.

Nutrient Deficiency Symptoms in Corn

(from: *Modern Corn Production,* 2nd ed., 1975, p. 256–266)

(Exact deficiencies may be difficult to identify because there may be more than one.)

Nitrogen—lower leaves yellow at tip and center, later dying; rest of plant pale yellowish green; slow growth; stalks slender; ears small, not filled at tip.

Phosphorus—leaves dark green, turning reddish-purple beginning at tips of upper leaves (in young plant; purple leaves in older plant usually from other causes [drought, barren stalk]); slow growth; stalk size small; ears small, misshaped, often twisted from missing rows of kernels, not filled at tip.

Potassium—lower leaves yellow at tip and edges, later dying; stalk weak; ears small, poorly filled at tip, kernels loose.

Magnesium—lower leaves streaked by yellow between the veins, sometimes with rows of dead spots; upper leaves may become reddish-purple at tip and edges.

Calcium—in young plant, leaf tips stick together (do not unfold), giving a ladder-like appearance.

Sulfur—yellow streaks between veins and stunted growth, especially in young plant.

Iron—upper leaves pale green to nearly white between veins (along entire leaf).

Copper—young leaves yellow as they emerge, becoming pale streaked between veins, edges may later die; youngest leaves twisted and dried; stalk soft and limp.

Boron—leaves brittle with small dead spots or streaks; top of growing plant with bushy appearance because stalk does not lengthen; tassels and ears reduced or do not emerge.

Manganese—leaves olive green, may become streaked; stalk may be thin and limber.

Zinc—leaves with wide whitish bands between edge and midrib (edges remain green or turn purplish) which may later die, youngest leaves may be white; plant short because of little stalk growth.

and waterlogging. Or sometimes the topsoil also may have poor tilth (structure). In general, adequate humus content will make soil loose and spongy, prevent crusting, and allow good drainage. Humus is produced when beneficial, aerobic (needing oxygen) soil organisms decompose organic matter such as manures and crop residues. However, to build up humus, the organic matter should have a high nitrogen content; animal manures or legumes can supply it. But if you plow under such materials into dense, tight soil, oxygen will be excluded and toxic fermentation will result. It is best to keep organic matter in the upper, aerobic zone of the soil, but gradually deepen it by plowing an inch deeper each year. Eventually a hardpan can be eliminated. An easy way to tell how deep your soil's aerobic zone is, is to dig up a corn or other plant's roots and see how far down the small feeder roots extend. Tilling in compost (already decomposed organic matter) in tight soils speeds up the process. Also, some problem soils are helped by plowing in soft rock phosphate and calcite lime, and in some cases, commercial soil conditioners can help loosen soil (in other cases they don't; test first). Leaving the land lie under a sod grass crop for a season will increase organic matter (from dead roots and plowing it in before planting corn) and increase stored moisture.

Trace element deficiencies can occur, but the actual cause is not always a true deficiency in the soil, but may be unavailability to the plant because of too low or too high pH or unbalanced major soil nutrients (such as not enough phosphorus to act as a carrier in the plant). For example, high potassium can lead to a magnesium deficiency. True deficiencies occur most often in sandy and highly weathered soils (in high rainfall climates). High lime soils are sometimes deficient in iron, zinc, manganese, and copper. The increased use of very pure manufactured fertilizers since the 1940s, the decreased use of animal manures, and increased plant populations have worsened trace element deficiencies (impure manufactured fertilizers and natural rock fertilizers have trace elements). A severe deficiency of any element will show up in the growing plant as certain symptoms (see box on previous page).

Correcting trace element deficiency or unavailability is tricky. Leaf symptoms and tissue analysis may indicate that the plant isn't getting enough, but don't indicate what the cause might be in the soil—whether a true deficiency, a tie-up from pH or nutrient imbalance, or even waterlogging or drought, which interfere with root absorption. It takes some careful detective work to find the cause(s). Foliar feeding can correct a deficiency in that year's crop, but is a band-aid treatment. Just throwing on a "shotgun" trace element mixture can often make the problem worse and make the soil toxic from trace elements it didn't need.

If soil is tight and short of humus, correcting that problem will sometimes correct trace element problems, since humus is a storehouse of trace elements in the proper forms, and the beneficial organisms living in it help feed nutrients to roots. Also, in soil that is deficient in major elements—nitrogen, phosphorus, potassium, calcium—trace element problems are magically solved when the proper balance of major elements is obtained. Available calcium should be high, and available phosphorus should be higher than potassium (many experts will question that, but if other soil conditions are correct, it holds true).

Corn is relatively sensitive to soluble salts, and thus is not well suited to saline or alkaline soils, irrigation with water of high salt content, and the use of concentrated high salt fertilizers (they can sometimes be used in lower amounts). High salt fertilizers include (in descending order of salt index): potassium chloride, ammonium nitrate, sodium nitrate, urea, potassium nitrate, and ammonium sulfate. Low salt materials include (in order of increasing salt index): superphosphate, calcium sulfate, triple superphosphate, mono-ammonium phosphate, diammonium phosphate, sulpo-mag (sulfate of potash-magnesia), and potassium sulfate. Salt fertilizers should be placed away from the seed ($1^1/_2$ to 2 inches below and to one side). Chloride-containing materials should always be avoided, since corn is moderately sensitive to chlorides, and soil bacteria much more so.

Weather. The final problem we will mention in this chapter is the weather, which seems to do just the opposite of what we want—rain when we want to plant or harvest and drought when the crop is trying to grow. Too high (over 90°F) or too low (under 50°F) temperatures are detrimental to growth and yield. Too much and too little soil moisture decrease the roots' ability to absorb water and nutrients. Too little light (cloudy weather) decreases photosynthesis and growth. Wind and hailstorms can decimate a growing crop.

You may think that you have no control over such problems, since you can't control the weather, but actually, the condition of your soil and the health of your crop determine how badly adverse weather affects the crop.

If the soil has plenty of humus and is loose and spongy, it will be well-drained but will soak up a tremendous amount of moisture and hold it for the plants during dry weather. You can drive around and see a lush green field next to one that is burned up by a drought; the difference is in the soil.

Also, healthy plants can withstand temperature extremes (including frost) and cloudy weather better than stressed ones. They also have stronger (but more elastic) stalks to withstand wind and can recover faster from all but the worst hail damage. Ammonium sulfate acts as a temperature moderator, warming soil in the spring and cooling it in the summer.

PROBLEMS? Sure, we all have them. Do we let them get us down or are they a challenge to overcome?

CHAPTER 5

Harvesting and Storage

How you go about planting and growing corn depends to a large extent on what you are going to do with it when you get done. Are you growing it for grain? To feed to your animals or to sell? Or are you going to make silage or green chop it? Or perhaps are you growing a specialty crop—sweetcorn for canning, or popcorn?

Each purpose for growing corn has its own set of limitations and desired goals. We will mainly concentrate on growing for grain and for silage (popcorn is harvested as grain, and sweetcorn is harvested early like silage).

Generally, when growing corn for grain, you will want to plant a variety that has a short enough growing season to mature before frost, and plant it early to be sure it gives a high yield. The use of a high quality seed and a seed treatment also helps. On the other hand, for silage, mature grain before frost is not necessary, so you can plant a longer season variety and higher population than for grain.

Silage. Ensiling is an excellent method of preserving a nutritious, palatable feed for dairy and feeder cattle, with little waste and easy mechanization. When moist whole corn or ear corn is chopped and packed into an air tight silo (or trench or tube), aerobic bacteria, yeasts, and fungi naturally occurring on the corn plants will begin fermenting the plant material. But in several hours they use up the oxygen they need to remain active, so anaerobic bacteria take over and continue decomposition, producing lactic and acetic acids as by-products. After about two weeks, conditions are too acid (pH 3.5 to 4.0) for them to remain active, so all microbial activity more or less stops and the silage is preserved. But if there are leaks in the silo or when you begin feeding the silage, oxygen enters and decomposition by bacteria and fungi continues, spoiling some of the silage. Poorly packed silage will also develop pockets of spoilage.

The commonly recommended stage of maturity and moisture for harvesting corn for silage is when the kernels have started to harden and glaze,* or when kernel moisture is from 30 to 35% moisture and the total chopped material is 60 to 70% moisture. Water can be added to nearly ripe corn to bring it to the proper moisture so it will pack well.

There are several materials that are sometimes added to silage for various purposes. (1) Lactobacillus inoculants serve to introduce the right kind of acid-producing bacteria, to insure correct preservation. Although such bacteria are supposed to occur on the corn naturally, the use of an inoculant can be cost effective. (2) Ammonia and urea are being added to increase the nitrogen content of the silage. It is said that ruminants can use the nitrogen to make extra protein; thus weight gain or milk production increases. This is a band-aid approach. The animals really don't need protein as much as high energy foods (carbohydrates, oils). High quality feed will have adequate protein already, or else a high quality supplement such as soybean oil meal should. If urea or ammonia are used, they must be balanced by a high energy content feed or else liver damage can occur. Urea can be toxic in too great amounts, and should be mixed thoroughly in the silage. (3) Gypsum (calcium sulfate) or sodium sulfate are used as a source of sulfur if urea is added; this should not be necessary if high quality feed is used. (4) Limestone is used in combination with urea to give increased gains, to neutralize strong acid additives (covered next), or else is added to lower the nitrate content of high nitrate feed. Again, high quality feed eliminates these needs. (5) Propionic acid or other chemical additives (sulfur dioxide, sodium metabisulfite, phosphoric acid, formic acid, sulfuric acid, hydrochloric acid) short-circuit or speed up the

---

*This is true of average corn, but with the high quality corn you can raise on good soil, ensiling can be done when the grain is in the milk stage. At that stage the calcium present in the grain is in a form more readily used by animals. Also, high quality grain will have a lower moisture content than average corn at the same stage of maturity.

attaining of acid conditions, preventing spoilage at high moisture levels. But the desirable partial breakdown of proteins that would occur by bacteria is also short-circuited. Such preservatives should be unnecessary for high sugar content crops (corn, sorghum, Sudan, small grains) and especially not for high quality crops. (6) Molasses, ground grain, and other carbohydrate sources, or whey are added to increase feed value and palatability, and at the same time to encourage growth of beneficial silage bacteria and speed up the ensiling process. Whey gives the least satisfactory results.

Nitrates. One serious problem that can arise when the whole corn plant is fed to animals is nitrate poisoning, which at high levels produces oxygen deficiency in the animals, evidenced by blue mucus membranes and chocolate-brown blood. Death follows. Milder sub-lethal cases include vitamin A deficiency (pink eye), thyroid disturbances (iodine deficiency), reproductive difficulties, lower milk production, reduced weight gain, and oxygen deficiency.

In a healthy corn plant, nitrates will be made into proteins and stored in the grain, but under certain conditions this does not occur normally and excess nitrates accumulate in lower leaves and especially the lower stalk. The conditions that favor high nitrates in corn stover are (1) high available nitrogen or potassium in the soil (from excessive nitrogen and potassium fertilization), (2) an extreme shortage of phosphorus or potassium (reduces growth while roots continue to take in nitrate nitrogen), (3) drought, (4) barren stalks (no kernels to store proteins in), (5) shading of leaves, either from high plant populations or cloudy weather, (6) a heavy rain after a long dry period (roots absorb much nitrate along with water), and (7) certain herbicides, especially 2,4-D (prevent nitrates from being made into proteins).

If you are forced to use high nitrate feed, here are some points to consider:

1. Non-ruminants (hogs and poultry) can tolerate more nitrates than ruminants (cattle, sheep). Feed the nitrate feed gradually throughout a day and increase the amount gradually. That way rumen bacteria can adjust and partly detoxify it. Adding grain, molasses, limestone, or live yeast culture to the feed reduces the effect

Corn Ethanol

For the last several years, the possibility of using corn to produce a vehicle fuel—ethanol—has generated much excitement among politicians and the agribusiness industry. Can growing our own fuel really free the U.S. from importing so much foreign oil?

Ethanol is just the chemist's name for what is often called ethyl alcohol or grain alcohol. It can and has been produced from many sources for thousands of years: fruits, grains, potatoes and sugar cane, for example. It has usually been made for human consumption in the form of beer, wines, vodka, whiskey and many others. Corn grain is just one convenient source.

All that is needed is an organic material (or biomass) containing a source of sugar (starch and cellulose can be changed into sugar). Certain yeasts can transform (ferment) sugar into alcohol. When corn is used, the left-over solids, called distiller's grain, can be used as livestock feed, and wastewater (high in nitrogen) can be used as a fertilizer.

With recent price increases in petroleum, interest in biofuels has greatly increased, with scores of ethanol plants going up, especially in corn belt states. As long as the base price for oil is high enough, ethanol production can be profitable, but in late 2008, the petroleum price nose-dived, and many ethanol plants went out of business. During the brief heyday of corn ethanol production (especially 2007-2008), the demand for corn was so high, with as much as 25 to 30% of the U.S. harvest used for ethanol, that livestock feeders had trouble obtaining corn, driving prices as high as $4 a bushel. Corn farmers loved that and increased their corn acreage (even plowing up land set aside for conservation), but it all caused considerable economic disruption. Critics questioned the wisdom of diverting a food crop to fuel when malnutrition and starvation are worldwide problems.

But besides the uncertain economic aspects of ethanol production, there are several other drawbacks. For one, the use of corn as a fuel source is hardly worthwhile, ecologically speaking. Depending on whose figures you use, growing and distilling corn may or may not take more energy than the ethanol produces when it is burned as fuel. You have to include the energy used to plant and harvest the crop, and to make the fertilizer, herbicide and pesticides that most corn farmers use (all of those products are usually made from and with petroleum or natural gas). Then the distillation process also uses a lot of fossil fuel.

Also, when ethanol is used as a vehicle fuel, it does not give as many miles per gallon as gasoline (only about 66% as many), so ethanol is blended with gasoline (another reason for this is that engines require special modification to burn pure ethanol). Burning ethanol does produce less carbon dioxide (CO_2) than an equal amount of gasoline, but if the use of petroleum in corn growing and distillation is added in, corn ethanol isn't a great help in fighting climate change.

Many critics of corn ethanol are hoping that the commercial production of ethanol from other sources of biomass, such as switchgrass, sugar beets, crop wastes, lumbering wastes, city leaves, garbage, and so on, will supplant the use of corn, These alternate sources would likely be more economical than using corn, but if some of the crops (switchgrass and other dryland plants) require tilling thousands of acres of marginal farmland, prairies and deserts, the ecological destruction and dust-bowl conditions could be devastating.

Although corn-based fuel may not turn out to be such a great thing, ethanol from other sources, as well as biodiesel made from oil seed crops (especially soybeans), waste vegetable oils and even single-celled algae may eventually help to alleviate the future shortage of petroleum.

of high nitrate (corn *grain* has little or no nitrate). Molasses and grain are high energy foods, giving the animals added strength.

2. Ensiling high nitrate corn reduces nitrates considerably after a few weeks. Do not feed green-chopped fodder, and cut the stalks high ($1^1/_2$ to 2 feet). Delay harvesting as long as possible, since nitrates decrease in time.

3. Also check for nitrates in the animals' drinking water. Nitrates in water are absorbed into the blood faster than from food, and they add to what is in the feed, making the problem worse.

Other methods. Green-chop fodder feeding can be an excellent source of nutritious animal feed if the crop is of high quality. In fact, high quality corn fodder can be a complete ration (if essential proteins, vitamins, minerals, and carbohydrates are present in adequate amounts). High quality corn can also be pelletized or baled like hay at the tassel stage or later. Mixing alfalfa or grass forages, straw, or grain with the corn reduces moisture and may improve feed value.

Corn for grain. The ideal goal in growing corn for grain is for the grain to completely fill with high levels of proteins, starches, oils, minerals, and vitamins; to become physiologically mature; to *grow* dry; and for the plant to stay standing until harvest. Some of those terms or concepts may be new to you, so let's explain them.

High quality corn will have its kernels packed full of nutrients, and as it matures, excess water will be "pushed out," and it will grow dry enough to store. Physiological maturity means the kernel has reached its highest dry weight, when no more food is being stored in the grain. It is indicated by a "black layer" that forms at the tip of the kernel. The plant should remain green until maturity or beyond. Early-dying plants, dented or shriveled kernels, low test weight, and high moisture grain all indicate low quality. Something went wrong—perhaps weather stress, low soil fertility in the latter part of the growing season, weeds, pests, diseases, etc.

In case you read over it, let's repeat: high quality corn should not dent (even "dent" varieties will dent little if at all) and should store without molding as it comes out of the field. That sounds hard to believe, I know. High quality corn may have a fairly high percent moisture (perhaps 24 or 28%), but this is "bound mois-

ture," which is locked into the stored food molecules. It should store without molding. If such corn is artificially dried down to 13% moisture like ordinary corn has to be, its food value will be greatly reduced. It should not be artificially dried. It will also have a high test weight, say 60 or 62 lbs. All of the above refers to truly high quality corn that has not been cut short by an early frost, not to just above average 56 lb. test weight corn, which probably will have to be dried to prevent mold. Sad to say, too much corn being grown these days is 52, 48, or 46 lb.

High moisture corn. Shelled corn or earcorn harvested early at 25–33% moisture and stored in an air-tight (sealed) silo is an excellent animal feed (if it is of high quality). If it is finely ground and firmly packed, high moisture corn can even be stored in conventional silos or trench silos. Storing high moisture corn eliminates any possible need for drying, but it also eliminates the possibility of selling the corn on the grain market. Preservation with propionic or propionic/acetic acids or sodium metabisulfite is often used to prevent mold; this should not be necessary with high quality corn.

Dry grain. For ordinary harvesting of dry earcorn or shelled corn, the kernel moisture content in the field should be from 21 to 28% for

average quality corn. As we have said, high quality corn may be as dry as it will get at a higher moisture content than average corn.

For storage of average quality earcorn, the kernel moisture should be below 23%, and below 14% for shelled corn. High quality corn will store without molding at higher moisture. Safe storage time is increased the colder the weather and the lower the kernel moisture content. Ear corn can be stored in cribs without the extra expense of ventilators and drying. Grain bins should be well designed and provide adequate ventilation and temperature control.

Grain drying has become a large, expensive, and seemingly necessary part of growing corn these days. But it is really necessary only for average or poor quality corn or to satisfy legal moisture requirements (which were written for poor quality corn). Drying with too high temperature literally cooks the grain, destroys its life, and greatly decreases its food value. If you must dry your corn, then you must, but it isn't always necessary to do it the expensive way, using precious fossil fuel. Field drying on the stalk works well as long as there is little lodging. Low temperature drying (with the air heated only several degrees above the outside temperature) takes longer but saves much fuel and doesn't harm the grain. It is hardly practical with large amounts of corn in cool, humid climates, however. Solar driers have recently been developed and can be home-made. Plans may be available from your state university extension service.

Mold. Whether in silage, high moisture corn, or dry grain, *molds are a ticking time bomb.* The potential danger is immense, and a number of disasters have already occurred. Some fungi are harmless and some, like *Penicillium,* produce materials useful to us. But other fungi produce poisons, called mycotoxins, which can be deadly. The best known mycotoxins are the aflatoxins, produced by *Aspergillus* fungi, common among the molds that attack poor quality and too-moist grain. Abnormal or unhealthy crops are most often attacked: cracked kernels of stored high moisture grain or crops damaged by hail or early freeze, or suffering stress from drought and insect damage.

Symptoms of poisoning in animals include impaired liver functions, reduced blood clotting, fragile capillaries and hemorrhaging, kidney damage, impaired immunity to diseases, interference with vitamins A and D and with calcium metabolism, anemia, reduced weight gain and production of milk or eggs, and poor reproduction.

An animal's milk, eggs, or meat can be contaminated by the toxins, and humans can also be exposed by directly eating food products made from moldy grain (cornmeal, grits, etc., also peanuts).

The common practice is to blend moldy grain with good grain to reduce toxins to "safe" levels before feeding to animals. You should be suspect of any ground feed you buy that wasn't made from your own grain. A black light screening test to detect molds should be made on all batches of grain or feed that you buy.

To prevent molds from starting in your corn, try to grow healthy, stress-free crops and store grain under cool, dry conditions. Chemical additives for high moisture grain to kill molds may be necessary if quality is not very high.

CHAPTER 6

Quality

A re high quality crops important to you? Is it really worth all the trouble it takes to grow better crops?

If you are feeding your crops to your animals, then you probably can see the value of feeding more nutritious feed and getting higher-producing and healthier animals. Even here, some of the "experts" tell us that it doesn't pay economically. That it is easier and cheaper to just give the animals vitamin and mineral supplements to make up for what *should* be in the crops—but isn't. But is it really better that way? Often the synthetic vitamins that are used in most supplements aren't able to be effectively utilized by your animals. Animals (and people) do better when they eat a well-balanced diet containing vitamins, minerals, enzymes, amino acids, proteins, etc., in natural forms. Even though the nutritionists and biochemists insist it makes no difference, a man-made feed mix or supplement can never have all of the literally hundreds of organic substances contained in the cells of healthy plants—not only vitamins, amino acids, and proteins, but also nucleic acids, nucleotides, peptides, amines, purines, phospholipids, enzymes, coenzymes, flavins, flavenoids, carotenoids, cytochromes, alkaloids, organic acids, polyphenols, and many others you have never heard of.

Even minerals such as calcium must be changed into organic forms by being complexed by a living plant or the microorganisms in an animal's digestive tract in order for best usage by the animal[1]. So in spite of what the experts say, animals are living organisms and function best when fed high quality, natural food.

[1] *Eco-Farm — An Acres U.S.A. Primer,* 1979, p. 398.

OK, you say—so high quality is important when you feed your crops, but I grow corn to sell, and the market only pays for bushels; why should I worry about quality? First of all, you aren't really paid by volume—bushels—but by weight —a 56 lb. bushel as a standard. The heavier your corn, the more profit you make.* High quality grain will automatically have a higher test weight, plus will not be docked because of cracked kernels or mold. Partly because of the large amount of horribly poor quality grain now being grown, the market is starting to pay a premium for high quality, especially foreign markets. Also, if you grow high quality grain, you should investigate specialty markets which will always pay a premium, such as poultry feed and bird seed companies, zoos, bakeries, and health food companies.

You can also come out ahead in the long run by striving for high quality crops because of the money you can save in grain drying, preservatives, fuel (good soil is loose, easier to work), and veterinary bills, to name a few.

How to tell quality. How can you recognize high quality corn, and how can you test your own corn for quality? We mentioned some of the signs of high quality in previous chapters: a deep, strong root system, vigorous growth, freedom from diseases and pests, early maturity with the plant staying green, well-filled cobs (more than one good cob per plant), high test weight, and kernels that do not mold in storage.

Another way of measuring crop quality as well as monitoring the plant's health while it is growing is to use a refractometer to measure sugar content. A refractometer is a precision optical instrument that allows you to quickly measure the percent sugars in the sap of a plant, which is correlated with the plant's food-producing efficiency (photosynthesis) as well as with food value—protein and mineral content. Refractometers are routinely used in the food industry, by canneries, wineries, and breweries for example, to measure the quality of the fruits and vegetables they buy from farmers or of the foods and drinks they manufacture.

---

* Test weight can be easily measured: take two quarts (dry measure) of corn and weigh in ounces. Subtract the weight of the container. The weight in ounces equals pounds per bushel. For example, if two quarts weigh 54 oz., the test weight is 54 lbs. per bu.

Brix Standards (% Sucrose) for Corn and Other Grains

	poor	average	good	excellent
corn, stalk	4	8	14	20
corn, young	6	10	18	24
sweetcorn	6	10	18	24
small grains	6	10	14	18
sorghum	6	10	22	30

Using a refractometer is easy. Simply squeeze a few drops of juice from the stems or leaves of the plant onto the glass prism of the refractometer, close the "lid," and look through the eyepiece. The sugar content is read on a numbered scale in units called Brix (same as percent). By comparing with standard levels (see table) or past readings that you have made, you can see how your crops measure up that day. You should realize, however, that the sugar content will vary—from one plant to another, from one part of the plant to another (higher in leaves and upper stalk), in different

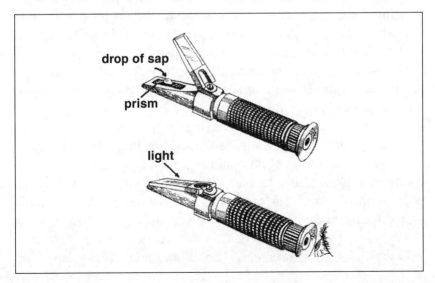

weather and times of day (higher on warm sunny days in the after-
noon), and for sick and healthy plants (sometimes a sick plant will
give a high reading if the sugar isn't being carried into the cob like
it should be). So you should be careful to check the sugar content
at the same place in the plant (say 3rd leaf-level of the stalk, or cob
level), and at the same time of day on sunny days.

What about standard lab tests for grain and feed quality? One
of the most commonly used tests, crude protein, tells you almost
nothing. It doesn't even measure protein, but only estimates it, by
multiplying the nitrogen content by 6.25. If there is a high amount of
nitrate or other non-protein nitrogen present, the result is meaning-
less. Other tests, such as digestible protein, total digestible nutrients
(TDN), crude fiber, and dry matter are also *estimates* of food value.
An amino acid analysis is another, more accurate, but very expensive test. All in all, the cheapest, quickest, and easiest food value test
is the sugar content reading made with a refractometer, since mineral content and truly useable protein content are directly correlated
with sugar content (unless the plant has a high sugar reading but is
obviously sick). Ordinary lab tests do not measure how well your
animals can utilize protein. A crop with a high protein test from a lab
but a low refractometer reading would not be as good for animals as
one with a high refractometer reading and a lower protein reading.

And then, the ultimate test of crop quality is how well your
animals do on it. They should give good gains and production, be
healthy, and reproduce well (if other factors are adequate, such as
water quality, ventilation, and environment). You can even use the
refractometer to test the animals' urine and immediately see how
they are doing on a certain batch of feed.

How to achieve quality. To summarize the methods of growing
top quality corn that are scattered through the preceding chapters:

1. Get your soil in good condition, with adequate humus and
balanced high fertility. Loose, well aerated soil with a healthy
population of beneficial organisms is essential for high quality
crops. Toxins make this difficult or impossible.

2. Select good quality seed of a variety that is well adapted to
your area. Consider climate, soil conditions and fertility, length of

growing season, desired use of corn (silage, grain to sell or feed), and hybrid vs. open pollinated (open pollinated gives better food value when grown on good soil).

3. Prepare a good seedbed. Plant when soil temperature is high enough, with appropriate population and row width.

4. Use starter fertilizer and be prepared to side dress during the growing season if soil conditions require. Use only those fertilizer materials that do not harm plants or soil life.

5. Monitor growing crops for problems: weeds, pests, diseases, and deficiencies. If possible, use non-toxic methods to attack problems.

6. Use appropriate harvesting and storage methods, depending on desired use of corn. Artificial preservatives, additives, and drying should not be necessary for high quality corn.

Improving your own open pollinated corn. If you want to try growing open pollinated corn, be prepared to go to some extra trouble and care. High quality often takes more work, but it is worth it. After you find an open pollinated variety you like, you can then grow your own seed and improve it year after year to obtain a variety that is "tailor made" for your climate and soil.

First, if you are not certain what characteristics you want (high protein, high mineral, high sugar, etc.), buy seed from more than one variety to test. Buy good quality seed with a high germination rate from a reputable company or individual. Some of the large seed companies carry open pollinated seed, as do some small companies and certain individuals. Plant test plots that are separated from one another to avoid cross pollination with other varieties. Be sure your soil is in top-notch condition. Go to extra trouble to control weeds and side dress needed fertilizers. Keep detailed records of all you do: the weather, crop growth and performance, and yield. Test grain for feed value, and if possible do animal feeding tests. Perhaps one variety will prove best for you, or perhaps mixing grain from two or three varieties will give you the best animal feed.

After you have decided on the best variety (or varieties), begin improving that variety by a selection process. There are several ways to improve varieties. The simplest is called mass selection. You just walk through the field (or hunt through the corn crib) and pick out the best looking plants or ears (don't use cobs from the outside dozen or so rows), collect them (take both ears of two-eared plants) and shell them, mix the kernels all together, and plant that seed next year. Discard kernels at both ends of the cob; those on the central several inches will be more uniform. Sometimes it takes a long time to show much improvement using mass selection.

Another selection process that can give more rapid results is called ear-to-row selection. Again, collect the best ears, but instead of mixing seed from different ears, keep each ear's seed separate and plant a separate row with each ear (this can be done with an ordinary planter if seed from each cob is put in a separate box). That way you can more rapidly increase the amount of seed from the best ears and eliminate the less desirable characteristics. The best ears from the best rows can be used for next year's seed. The experimental rows should be in a larger field of the same variety to eliminate stray pollen. To prevent inbreeding in the ear-to-row method, it is best to detassel alternate rows and save seed only from detasseled rows. It would help to record any notable characteristics of each ear, number them, and label each row.

In either method of selection, choose the ears or plants that show the characteristics you want to increase in the next generation, such as height of plant; number of ears; resistance to lodging, disease, or pests; early maturity; ear or kernel size; protein, oil, mineral, or sugar content; yield; etc. But remember that the open pollinated varieties have much genetic variation, not found in commercial hybrids. This genetic variablility is good because it allows the crop to do well in many kinds of conditions. Therefore, don't expect all plants in the field to look alike, and don't necessarily choose cobs from identical-looking plants to use for next year's seed. Instead, focus on the particular traits you wish to increase. It may take several years to get to a desirable level.

After you have developed one or more varieties of your own, you may then want to try variety hybridization to obtain still new varieties of open pollinated corn which may have much improved characteristics. Either two varieties can be planted in alternating rows and be allowed to cross pollinate at random, or one variety can be detasseled and serve as the female parent. The best ears or kernels from detasseled plants can be planted next year. Not all offspring from crosses will have the desired characteristics, but selection over several years may develop a variety that will breed true. Some of the best open pollinated varieties, including the excellent Reid Yellow Dent variety, originated as variety hybrids.

If you engage in growing your own seed corn, you need to take precautions to insure proper storage and good germination. Let the selected ears mature and ripen in the field, but do not let seed be exposed to hard freezes. Select cobs that show the same degree of maturity (all early or all medium). Cobs or seed should be stored in a cool, dry place, away from rodents and stored grain insect pests. Storage at 40–50°F and in a room with a humidity below 13% is best. Cobs or seed should be laid out so air can circulate. Seed shelled from cobs for ear-to-row planting can be kept in small paper sacks that are left open at the top. If seed corn has to be artificially dried, it should be done slowly at low temperatures (below 107°F). Germination rate should be tested by wrapping about 50 (or more) seeds in a moist paper towel at room temperature. If stored grain pests infest the seed corn (weevils, grain beetles, grain moths), they can be killed without much harm to the seed if the seed is heated to 140°F for 10 minutes. Avoid cracking kernels if a mechanical sheller is used.

So that's how *you* can grow top quality corn. It takes extra work. It's more trouble. You have to do more thinking. Is it worth it? I think so.

Index

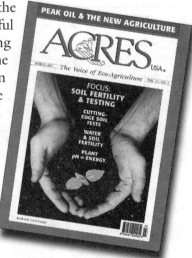